HENRY M. BURTON

Your First Research Paper

Learn how to start, structure, write and publish a perfect research paper to get the top mark

First edition

ISBN: 979-8-5530-9521-5

This book was professionally typeset on Reedsy.
Find out more at reedsy.com

Contents

Introduction

So, it's time for you to sit down and write a well-thought-out paper, is it? Are you worried about not getting the details just right? Perhaps you are concerned that your writing is not up to snuff, or you do not know where to start first. However, writing a paper does not have to be difficult or scary. It may take time, especially if you want to go through the process of researching everything to make sure that all details are perfect. It is easy to get stuck in front of your screen, staring at that blinking cursor on it and wondering what comes next for you or how you can make sure that you do, in fact, succeed when it comes to providing yourself with what you want. Nevertheless, do not forget this one important point: Writing a paper is not about getting every single word phrased correctly the first time that you write it. It comes in steps that take time and effort. Forget that notion that the first draft of your paper has to be perfect. Forget the idea that you must write the final draft the first time fingers meet keys. Writing is fluid. It is a process. It takes time.

This book is here to guide you through the art of writing a proper research paper. If you are working toward completing your first formal paper for your field, no matter where you are in your degree-seeking process, this is the book for you. Did you just complete your first lab and need to get the results written? Perfect! Follow this guide. Did you just finish up a marketing research project? This can help you with that. Maybe you just studied something for psychology, and you need to write up your report—and as you may have guessed it, this book is for you! No matter the reason behind the paper, you can learn to write it with ease with the information you are about to be given. As you read through this paper, you will see everything that matters the most when it comes to writing a paper, and you will be guided through every step of the process. You can learn how to properly write anything so long as you follow these steps. Now, don't let that fear or that anticipation of not being able to get the details right hold you back. It is time to get started!

1

Introduction to Research Papers

What events led up to the Great Depression? Why do people want to live in social groups? What makes us unique as a species? How do we think? What goes into writing a paper? These are all questions that could be posed. They are all very important in their own ways, and they all share one thing in common. Despite how different these questions all look, the one thing that they do share in common is that they all have answers to them. The answers may be different, but they all have factual answers that could be found if you were to research them online. This means that they are problems that can be solved. They are mysteries that already have an ending to them. All you have to do is go out and find those answers for yourself.

We all have questions that we either want to answer, have to answer or are prompted to answer throughout life. We have questions daily—what temperature do I cook this food at? What is the weather going to be like this week? Where should I invest my money? What career is the best one to get into right now? Thoughts with questions pop into mind constantly, and our natural reaction is to look up the answers. However, sometimes, we have more formal questions presented to us at varying points in time. If you are in college or university, you are probably well acquainted with the idea of having to write papers, but oftentimes, people find themselves woefully unprepared when it comes to an understanding of how to create well-crafted research papers. However, they are very important to learn how to use if you have any intention of pursuing academics.

Within this chapter, we are going to take a look at precisely what a research paper is. You need to figure out precisely what goes into it and how you can ensure that you can manage to create one the right way.

What is a Research Paper?

Let's begin by considering what research itself is. We research something when we have a question that needs answering. It could be that you want to know who built the pyramids in Egypt, or maybe someone asked you how a diesel engine works. You may also have questions that are presented to you to test your abilities to research and learn, such as being asked whether or not you understand what you are learning in class in college. Perhaps you are told that you must answer a question using the information that you have been provided over time. No matter the reason behind the question, the question always comes first, and you must remember to answer it at the end of the day. This involves research.

Most of us push our research to the internet these days. There is no reason to run out to the library to pick up books to read about the subject that matters to you unless you cannot get access to the internet in the first place. We type in our questions into our web browsers and read the results to understand what is going on, what matters the most to us, and how we can put that information to good use for ourselves. That is research, and it is done when you have a question that requires an answer.

A research paper, then, is meant to organize everything for you. It is the process through which you can present that question that you have for answering and then dictate it all out, explaining the trains of thought, the research done, and how you eventually arrived at an answer that you have. When you do this, you are essentially discovering how you can convey the information that you have gained throughout the period of time that you have researched.

The Purpose of the Research Paper

The research paper, then, has one very important purpose. No, it is not just to get a good grade in school, though it can help you achieve that. No, it is not just for doctorate students trying to get their Ph.D., though it can help them with that as well. Not only does it help you organize all of your information in one place, but it also ensures that at the end of the day, you will be able to communicate. It is great that all of your information will be contained in one place, but there is so much more to it than just being able to see it all written down where it matters the most. There is much more to that, at the end of the day. You must also be able to communicate what it was that is written.

The idea behind creating a research paper is that you have a structure that will help you to convey what it is that you have written at any point in time. You must be able to convey how you answered the question, as well as what your question was, to begin with. When you make use of a research paper, you can figure out precisely how you can communicate that information to others that may be interested in it as well.

Because research papers are structured, we can know what to expect within them. We know what is likely to be present and what will not be. We know that we can expect there to be information about the thesis that is being discussed, as well as what to expect with everything else as well. When you learn how to understand these research papers, you can learn to format everything into that standard format that will allow for the research to be followed and understood, and that is greatly important.

Ultimately, the purpose of the research paper is to present a thesis—the statement or theory that you have as to why something works. It discusses the research that you have gathered to support that thesis, for example. It will help you to narrow the scope of your research so that you can stay on track and on topic. It helps to guide your thoughts and keep your writing structured in a logical way. It helps you to create that narration of your understanding, your experience in researching, and everything else that you have come to understand through the time that you studied whatever the question that you had was in the first place.

Even better, because your results will be listed in a way that is standardized, organized, and able to be followed easily, you can present your argument to the world. Anyone can pick up a well-formed research paper and understand what the point of it was. Anyone can look at that research paper and pick out the parts that are there. Because of that, the research paper is highly valuable in many academic or research-based settings. That standardization is essential if you need to be able to share research readily, and that means that being able to create your own research paper, whether for your own personal benefit, for a job, or for school, is highly useful.

Think of the research paper as the ultimate way in which you can present the information that matters the most. It can help you explain the information that is relevant to the question that you had at hand, and it can present it in a standard format that will help you get your point across effectively, succinctly, and in a logical manner.

2

The Building Blocks of a Research Paper

When it comes right down to it, any good research paper has eight key sections that go into them, with the summary and conclusion being closely related to them. It is imperative that you understand that these key building blocks should be present in most research papers, though some fields will have their own preferences and expectations. It is imperative that you can understand each and every one of these blocks that go into the research paper, as if you find that you are missing any, you may find that your paper as a whole will suffer for it.

Your paper components are there for a reason, and they are designed to help you structure your paper into something specific—it is supposed to ensure that you have all of the information that you will need to understand the question that was researched and all of the data that went into it. As you go through these key sections, you will see the ways in which each and every one of them plays their own part in ensuring that the paper is as effective as possible. In a research paper, it is imperative that all information that you provide be for a very specific purpose, and that is why you do get this sort of template that you can rely on to ensure that your paper is as effective as it can be.

Abstract

The first part of your paper, after the title page, is going to be your abstract. It has a very specific function to it, and you will need to make sure that you address all four key points within the abstract section. Keep in mind that your abstract is designed to be short—in most journals, it is between 200-300 words and is meant to provide all of the pertinent information to what the paper's purpose is, how it can be understood, what goes into it, and how you intend to answer it. It will summarize up the introduction, methods, results, and discussion that you will be presenting throughout the paper. In particular, it will answer the following questions within that one paragraph:

- **What question or questions were you researching, or what was the purpose of your research?**
  This is the key point of your introduction section and should be answered within the first sentence or two.

- **What was the design, and what methods did you use in your research?**
  This is the summary of the methods section and should explain how you have designed the study, as well as the methodology that you are using—briefly stated. You will have the entire methods section to go into detail with this later.

- **What were the key results or findings that you found—i.e., the answer to your question?**
  This is the summary of the results section and you should be able to report results and any relevant changes or trends that you found.

- **What was a summary of your conclusion?**
  This is the summary of discussion and should be able to clearly, but

succinctly state the results that you found through your research process.

This should all be done shortly with an active voice in past-tense. While it may come first in the paper, it is typically done last, allowing you to reference the details that you provided in the rest of the paper. Do not forget that you double-check that everything included in the abstract corroborates with the paper, and ensure that you only provide information in the abstract that will also come up in the results.

Introduction Section

The introduction is the point in which you are able to begin establishing context. It is not supposed to be too long, but it should ensure that any relevant research and citations make it in at this point. It should also provide the reader with an understanding of the topic at hand. It should also work to present your question that you are having as well as the thesis that you are studying.

This part of your paper will answer a few other key questions before it is completed:

- What was studied?
- Why did that particular topic of study matter?
- What was known prior to the study/research?
- How did the research/study improve upon/advance knowledge known about the topic?

This section ought to be presented in an active voice as well, and you should attempt to avoid first-person writing, but if it must be included or there is no natural way around it, it is acceptable to use some.

The structure of this is perhaps the most important part. When you present the structure of your introduction section, you should think broadly and slowly narrow it back down to the innermost point—the most detailed section of it all. To properly format your introduction, include the following in this order:

1. **The area or subject of interest:**
   This should involve keywords discussing and identifying what it is that is being addressed. If you have a research paper about where the Pyramids came from, you would probably want to make sure that Egyptian and Pyramids made it in within the first sentence or two.

2. **The context of the research:**
   Here, you want to summarize what is known prior to research and studies. This is done with a review of the research itself—you should discuss the information and cite them, but it should not be very long or detailed—that happens later. You would discuss, for example, what the pyramids are, where they are found, and theories on how they may have been created—broadly speaking, of course.

3. **The hypothesis investigated:**
   This is done quite clearly, especially early on—you can literally write out, "The intent/purpose of this research/study was…" or something similar in order to ensure that you are able to properly clarify what will be discussed.

4. **A clear statement on how you approached the problem:**
   Finally, the last part of your introduction should clearly but briefly tell what your approach to the research was done. What is it that you experimented with? Where did you go to understand the question at hand to support your hypothesis?

Methods Section

You may also see this section referred to as methods and materials some-times—it is designed to explain precisely how the study was done. Of course, this will vary somewhat based on the field that you are studying or writing your paper on. A scientific research paper will have you going into details such as what was studied, where you studied it, how you studied it, and how you collected your data.

This part of your paper may work well with subheadings that will help you see what goes where so you can properly keep track of everything. Keep in mind that this is not meant to be a step by step guide in the way that you may see it in a lab manual, but it should be clear enough and detailed enough that someone could replicate it easily from the way in which you provided. This means that you must include any relevant measurements during this point of your paper.

Make sure that the details here are sufficient, but not written out to be overly wordy. You want to write this in the least amount of words possible without losing meaning or content.

Results Section

The results section of your paper is where you finally begin to objectively report what you found. The point here, however, is that you are providing the results without any attempt to discuss or interpret them—that comes later. This should be done logically and in a manner that makes sense. For most research, you will probably want to use a combination of both text and graphs, tables, or other figures. You will always begin the process with text, but you will also make use of figures when relevant. If relevant, make sure that all axes of your graphs are labeled accordingly and that you provide scales that will enable them to be read and understood clearly.

This should be concise and objective with a more passive voice than before, but you will want to try to stick to active if possible. Make sure that, as with before, you always include past tense here—you are reporting studies on something that has already happened and thus deserves the past tense treatment. You should also make sure that you avoid anything that may even hint at interpretive, which can be easy to unintentionally add in this section.

As you write this, remember these considerations as well:

- **What were your results?**
  You want to figure out what it was that the results said and whether they were provided in a manner in which they can answer your question or hypothesis that you had earlier—whether a yes or no.

- **How are they reported?**
  You want to ensure that when you are reporting the results that you have found that you provide as much information as you can during this. You want to figure out whether or not there are any factors that may be conflicting or contradictory—if you have several groups that are being researched, you want to try to explain any factors that may have changed your results. Were you studying only male dogs or male and female when you wanted to know how high they could jump? Were you looking at whether or not they were different breeds? What else were you considering here? If there are any potential differences that ought to be reported, make sure that you do.

- **Organize your results:**
  You want to ensure that your information is organized, and if you do

include a Tables and Figure section, you want to ensure that they are organized in a way that makes sense while also ensuring that you properly label them and provide legends. The legends should be above tables and underneath figures.

- **Present negative results as well:**
Even if your research does not line up with the hypothesis, it should still be provided to you. This is something that may require you to change the way in which you approach what you are talking about, or it could also tell you that you were on the right track but need to change things. No matter the results, they deserve to be reported.

Discussion Section

The discussion point of your paper is there to help interpret. Remember, the results were all about simply providing those details, and now, you are tasked with interpreting those results, so you know what to expect and how to expect them. This should not only talk about what you already knew beforehand and putting that into context with your research, but it should also explain how you understand the information now that you have finished the research that you were doing. Again, there are some key questions that you will need to ask yourself to complete this section:

- Were your results able to answer the hypotheses? Yes or no? If yes, how can you interpret them?

- Do your results agree with past known data? If not, is there an obvious explanation, or did your (or their) experiment have a flaw somewhere?

- How do you now understand the problem that you sought to answer in your introduction?

• What is the next step in your research process?

These are all very important details to keep in mind—without them, you are bound to run into all sorts of structural problems. You want to make sure that you are concisely answering all of these questions as you format your paper, and you should make sure that your speech and writing styles make sense. Keep them clear, and while it is acceptable to use first-person speech, make sure that you are also not overdoing it.

As one final point to remember, you should never include any new information here—there should be no new results that did not first appear in the results section of your paper. Rather, you should only be interpreting the information here.

Summary and Conclusion Section

These sections are really just there to summarize. They are much shorter and less detailed than the previous sections, but they are highly important as well. You want to ensure that you meet several key points as you conclude your paper, including:

1. **Reiterate your topic:**
   Once again, you should briefly bring up the topic that is being provided—it should be no more than a single sentence for an effective paper.

2. **Restate your thesis/hypothesis:**
   This is also short—it should be narrowed and specific. Reword it so that it also works well with the reiteration of your topic at this point. It should flow well.

3. **Restate the main points:**
   At this point, you want to reiterate the main points of your paper—these are usually the topic sentences that you will find in all of the major paragraphs or sections. Make sure that you are only restating the point and not the supporting details of the points. For example, you would write, "This species of a dog jumped 5.2 inches higher than that species" and leave it at that without getting into details of how that is supported.

4. **Summarize it all up:** This is not always required, but it can help—you are able to summarize how all of those main points come together.

5. **Call to action—if necessary:**
   Sometimes, you can make use of a call to action, but only if you really need to address something important or that more research or effort needs to go into it.

6. **"What now?"**
   At this point, you then express what comes next—you are able to write out how the topic of your paper really matters.

Reference List

Finally, your paper should include a reference list in which you are able to provide all of the sources that you used. They may be online sources or others—all that matters is that you report anything that you referenced or anything that was cited within your paper. Ultimately, there are many different ways in which you can do this—it will vary greatly, and you will need to make use of the format that is standard for the field that you are in. The most common is the MLA (Modern Language Association) format and the APA (American Psychological Association) format. Each of these are slightly different from each other, and they will be specific to the field that you are in.

These will provide a few key details for the information that you have referenced so that you can provide that research elsewhere for yourself as well. They will provide:

- The date of publication and access
- The author's name
- The page numbers (if relevant)
- The title and publishing location

The most common structure is author first (last name, first name format), the title of the book second, including the full reference and how it is provided. Web URLs are commonly provided as well.

3

What Makes a Research Paper Effective?

If you have to write a research paper, you know that it has to be effective. After all, why else would it be so structured and regulated? Making an effective research paper can be intimidating; it can be scary to feel that pressure of having to figure out precisely how to word and format everything so that you can get your point across without wasting precious time elsewhere. If you want to do this, you must know what to focus on, and thankfully, you will have this guide here to help.

Figuring out how to make your research paper effective is as simple as understanding the true purpose of what you are setting out to do. You want to get in, get the job done as effectively as possible, and get out without wasting any time or any words. You do not want to bore your readers, nor do you want to leave them wondering why you did not provide enough depth or insight into what you have written. If you want to make your research paper effective, there are three key points to consider.

First, you want to make sure that you have an effective research question AND make sure that you also make that question as apparent as possible. Then, you want to show that you have done the work to answer that research question so that you can test your hypothesis about it. Finally, you want to provide your answer. If you make sure that you can do those three steps, you will be able to ensure that your paper is, at the very least, serving the purpose of what it sets out to do. From there, all you will have left to do will be to figure out how to flesh it all out in the most effective way possible.

The Research Question

The research question is the starting point of any good research paper, and if you do not have a solid research question, the rest of the paper will be lackluster as well. For this reason, in particular, you will find that you really want to focus on this point first. You will always want to make sure that your research question is effective and solid so that everything else can fall together nicely. Think of this as your paper's foundation. If your foundation is weak, everything else will fail as well. Let's take a look at what your research question must be.

Sometimes, your research question will be provided for you; you will not have to do anything to make it. This is especially true if you have been given a topic by a professor to explore. However, if you do have to make your own research question, you must make sure that it will work for the entirety of a research paper. This will require you to come up with a question that will meet the following standards:

- **Clear**:
  You should have a question that is clear and readily understood without being loaded with all sorts of jargon. This means that anyone in the target audience, regardless of their experience, should be able to get the gist of what you are exploring.

- **Focused:**
  You must have a narrow enough scope that you are able to answer the

question completely within the duration of your paper. Remember, this is a paper, not a book.

- **Concise**:
  You want to make sure that you use up the least amount of space possible so that you are not losing track of what matters or getting off-topic.

- **Complex**:
  Your answer should never be a simple yes or no at the end of it. For example, "Did the Egyptians build the pyramids?" as a blanket question is probably too simplistic of a question to get into, but it could be made more complex by asking, "How could the Egyptians have been able to build the pyramids?" Notice how, with that second question, you would have to go in and analyze the information about what is being explored. You would have to figure out whether they had the math and the technology to be able to make such perfect shapes, and that could be something that you could research.

- **Arguable**:
  This is where things get complicated. You must make sure that you are answering something that is potentially arguable or debatable. The idea is that you must be able to defend your position with the research that you are going to do.

When you can manage to hit all of those points, you know that you have the right question. However, making sure that you form the right question may be a bit more nuanced than you intend for. To get the right question, go through these steps:

- **Identify the general topic:**
  For the most part, you should be able to write research papers on topics

that you genuinely care about. Start broadly, and you will be able to narrow it down over time. For example, you may start out with "Ancient Egypt" on your way to figuring out what your proper research topic will be.

- **Do general preliminary research:**
Unless you are already familiar with the topic, you should begin with some quick searches to determine the current research being done. This will also help you figure out if you have any other questions that you may have about the topic.

- **Consider your audience:**
Is this a paper for your degree? Is it for a job? Figure out what the audience is and keep them in mind. Do people in your intended audience care about the question that you are thinking about?

- **Brainstorm:**
Start writing down any questions that you come up with. Do not feel like you need to figure out the perfect question right this moment. If you think of it, write it down. You can eliminate it later.

- **Evaluate:**
Now, you can start thinking about the questions themselves. Evaluate all of your potential questions based on the standards listed above. If any of them do not meet all five points, eliminate them, or alter them to make them fit. Then, choose your favorite one.

- **Research:**
At this point, it is time to start considering how to get moving on your question. If you determine that you can, in fact, do the research with the information and materials that you have available to you, then you have probably chosen a good question. If you find that research is too difficult, it may be worth going back and choosing another topic instead.

Searching for the Answer to the Research Question with Methods Section

You must also make sure that, in an effective research paper, you take the time to do the research and organize it properly. Many of these different research methods will make use of data or experiments that are presented, such as in a science research paper. This section must provide enough details so that your research or experiment can be replicated.

Within this section, there are a few key points to remember as well that will keep you working toward what you need. You will want to remember to keep in mind the following two key points:

- **Keep it past tense:**
  You have already done the research. You must present your information in the methods section in the past tense.

- **Detailed but brief:**
  You need to tow that line between providing enough information that your research can be replicated, but you want to also leave enough room for discussion in the next few sections.

Your methods section will probably have a few key points within it. Of course, the particular sections that you will include will be dependent upon the scope of your study and how you designed your research or your experiment. In general, you can plan on having the following four subsections:

- **Participants:**
  If you have completed an experiment, you probably have some

participants, especially if it is a sociology or psychology paper. You will want to list out exactly who your participants were, how they were chosen, and anything that makes them unique as a group. Note if you chose randomly or if you had some criteria that went into it.

- **Materials:**
Anything that you used during your experiment or research needs to go here. This includes equipment, books, websites, pictures, graphs, tools, stimuli, or anything else that was possibly used during your time. If you used something specialized, you should probably include information about that as well.

- **Design:**
This step will help you discuss how you came up with your experiment or research and how you were able to define it. Point out any variables, independent and dependent, controls, and anything else.

- **Procedure:**
Finally, this part will detail the way in which you engaged with the material and the experiment. Make sure that you are providing yourself with everything that goes into the process, but again—avoid being too wordy! If you provide too many details, you will probably lose your audience or take away from the impact. Consider the difference between "The research was done using these methods on these subjects with this material," versus, "The research was very carefully executed, facilitated, and cultivated through various means, including taking my eyes and tracking the words across the pages for every detail instead of just skimming over it. Every paragraph was read exactly three times before moving on to the next." That second example gave way more information than is necessary or helpful and only serves to distract from the material itself. This is not the time to impress your readers with your flowery purple prose—instead, keep it as succinct as you can.

Answering the Question with the Results Section

Finally, you must answer the question effectively as well in your results section. This is what really brings everything together. If you do not take the time to carefully answer your question, you will actually miss the entire point of your research paper—which was to answer the question. However, it is not enough to simply state what the answer to your question was. It is time for you to justify what happened. While the methods section will help you to come up with an organization to your experiment, the results section is necessary to justify everything. To make an effective results section, you will need to do the following:

- **Results should justify, or in some cases, contradict your claims:**
  You want to report your data honestly, and sometimes, it turns out that we were on the wrong page altogether with what we were researching—let your results section do that. It should simply report what you got out of your research. You can extrapolate from that later.

- **Summarize:**
  You want to offer a summary of the results—not the raw data. You should be writing them in a quick summary, and it should be brief. It does not have to present each and every part of the numbers and how they worked out.

- **Remember your tables and figures:**
  You want to make sure that you have the tables and figures there to make sure that data can be glanced at quickly. This allows you to avoid having to write out the results—because they are all displayed, and your reader will not have to read to get them. Start with your figures and then write around them.

- **Report statistics (if relevant):**
  If you need to make use of statistics, and you probably will, you need to worry just about reporting them. You do not need to explain any relevant statistical models—assume that someone reading your particular paper is already familiar with anything that you would be discussing.

In terms of formatting, keep the following points in mind as well:

- It should always be written in the past tense—all results were already determined.

- Make it objective—no interpretation here.

- Make sure results follow your formatting standard.

4

Step-by-Step Process

Now, you know what makes a research paper effective. You know what the format is. Now it is time to get started and get writing once and for all. It is time for you to put everything together so that you know what to expect and how to make it work for you. Let's go over those key points that you will need to consider to ensure that your paper is well-polished and ready to be presented right on time—every time.

Brainstorming and Researching the Topics

You begin with the brainstorming and research stage. Think about all of the topics within the scope of what you are going to be researching. Write down all of the ones that are interested and take some time looking into them. This is the process that you saw in the last chapter to provide you with everything that you would need to make an effective research question.

Stating the Thesis

When you have your research question chosen out, it is time to state your thesis. Your thesis statement will summarize what your point that you are trying to prove is—it will also briefly support the reason that you think this way. It is like your road map to figure out what will be found within the paper. Start by writing out your thesis before you really begin to outline your paper—this will help you to figure out what the structure for the rest of your paper will look like.

A solid thesis statement will make a blend out of the following points to create one or two sentences about your paper:

- It must state the topic

- It must state what you think about the topic

- It must offer support for your thoughts on the topic

- It should offer an objection to your thought, if applicable

- It should offer a counterargument to the objection that supports your thoughts on the topic

For example, you may have a structure that looks something like this:

*While [**objection to your thesis**], [**what you think the thesis is/counterargument to objection**] because [**support for your thesis**],* *[support for your thesis], and [support for your thesis].*

This is quite simple to write out when you follow this particular format, adjusting for grammar, you will find that it is easier than ever to get started and moving with everything.

Drafting the Outline

When it comes to drafting your outline, you will want to start vague. Write out what all of your sections and subsections will be and start to organize it. You should now have your thesis statement with your supporting evidence, and that is great—you want to make sure that you know what you are doing and how you are doing it. However, you must also make sure that you are figuring out the right way in which you can fill in the blanks as well.

Outline your arguments, work on adding anything that you think is relevant during this stage, and make sure that you provide everything that you think you will need during this stage. This should involve you starting to outline what you think about your research, putting it in order, and preparing it.

Refining Your Outline

With all of your outline written out, it is time for you to start parsing it down. If it is not directly relevant to the thesis or if it does not help you further your paper, cut it out. Start putting all of your writing and outlining into a logical sequence that will help you figure out how to structure everything and prepare to get started with your paper. You should be able to glance over your outline and have direction on where you will be going with your proper writing. You want to begin putting it in the right format, which you have been handed several times at this point in time in the book.

Keep in mind that your outline does not lock you down. It should be your guide, but if you need to veer off from it sometimes, that is okay. As long as you are able to keep everything in line with answering your research question and if you can justify the inclusion of something as support, then do so. It will be worth it in the long run.

Mapping Notes

When you have the general format down, or if you are struggling to make sure that your outline actually is on track, you may want to make use of a map. You can visually display the path of your paper if you go through the use of a mind map, which will help you see how everything relates together. If you are still unsure about some points in your paper, try mapping it out. Start with your topic in the center, and then make lines off from your topic for each of your key argument points. Then, put in all of the supporting evidence that you have. If, at the end of the process, you have evidence or support that does not naturally fit in anywhere, eliminate it. It was not necessary or essential to proving your point, so do not waste your time with it.

Avoiding Writer's Block

If you find that you are stuck with writer's block—every writer's worst nightmare—try considering a few key tips that you can use to jump-start the process. First, start with setting blocks of time for yourself. Perhaps you start small—dedicate yourself to writing for just five minutes. Tell yourself that you cannot get up until you have finished writing for those five minutes and then push the point. Make sure that you write without stopping to think about what you are writing. You can edit it the second pass around. Just get out your writing and get it out as quickly as you can. This is great to make sure that you are able to properly get the words out there. Once they are on the page, editing them is so much easier.

Another common tip is to even neglect typos as you type. They can distract you and make you go back, forgetting the train of thought that you were on in the first place. Write first and polish it up later. After all, your documents will have plenty of red and blue squiggle lines to catch your attention if there are any mistakes. Just get writing. That's the best way that you can avoid the trap of writer's block. It does not have to be perfect—it does not even have to be good or perfectly grammatically correct at this stage. It just has to be, and from there, you can polish it up.

The First Draft

The first draft of your document will be made while you make use of the above ideas. Just get it done and get it out there. It is incredibly rare for you to have a document that is entirely perfect the first time around; you must make sure to give credence to that and make it a point to figure out how best to fix anything that has problems.

Remember, you do not have to write your paper in order. It can help to do so, but if you have a well-designed outline, you should be able to jump around if that is easier for you. You may find that one area in your paper is particularly inspiring to you, and that is just fine—you can head over to that area and get writing. You can make it neater later. Just get it out there and be proud of yourself. That first draft is the hardest part! Once you have it, the polishing up of the document will be far easier than anything else. All you have to do is begin and make good use out of it.

Citing and Listing References

As a note, when you are doing your first draft, it can be distracting to write out all of your citations as you go. Instead, leave yourself a note, either in the document's note-taking capabilities, or on the side for you if you are not sure what it is that you need, or if you are in the flow of writing and do not want to be interrupted. Highlight any references and note the source briefly here—you can fill out the citations later, but they will need to get done.

Keep in mind that if you do not add citations, you will be plagiarizing. You must make sure that you give credit where credit is due, and if you use something in your paper, you must provide that. In particular, you will need to cite something when:

- You quote it
- You are generally paraphrasing something that someone else wrote
- Any data that is added to your paper
- Any images or graphs or charts that you use

Sometimes, you do not need a source at all—if you are using common knowledge, there is no reason to cite it. For example, you do not need to provide citations to the concept that there are pyramids in Egypt—that is common knowledge. However, if you get information about how people built them, you will probably want to cite that.

When you are citing your sources, make sure that you are addressing the guide for the style that is particular to your paper or your field. For example, most psychology papers will be cited in APA. Other social sciences also usually prefer this method, including sociology, political science, and education. When it comes to humanities, you will most likely need to make use of the

MLA format. Science has its own expectations for citations, as do medical journals. You will need to look at what is standard for your particular subject and seek out that style guide to make sure that you are, in fact, citing everything appropriately.

Review the First Draft

Next comes reviewing your first draft. Take some time and go over your paper, usually a day or two after your first draft was completed. This gives you time to let go of the information in your mind so that you can read it with fresh eyes that will catch any of the typos that may be present within it. Go over it and start to polish it up. You will likely have plenty of polishing up to do if your first draft was nothing but rushing through everything to get it onto the paper, but that is okay—just take some time and feel proud of yourself for accomplishing such a monumental feat. Check for typos, flow, continuity, and general structure. Fact check anything that you think you may have gotten wrong as well.

Peer Reviews

Though you will not always have the opportunity to do this, make use of a peer reviewer whenever possible. They will be able to help put fresh eyes onto your paper, and that can help figure out if there are any particular weaknesses that need to be ironed out to get everything in working order. Make sure that you do not get too upset or offended if someone has comments that you did not expect—this is part of the polishing process. Don't forget—coal turns into a diamond when under the highest pressure, and likewise, a well-scrutinized paper is likely to be far better formatted and written than one that has only been written and looked over once before submitting it.

Finalizing the Paper

After the peer review is complete, go over and make any of the recommended changes, if you find that they make sense. Take the feedback graciously and make sure that you take it all into consideration to figure out if it is right for you or not. Make sure that you take advantage of this process and work through it to the best of your ability. This is the last stage of any polishing that you are going to do.

After you have finished making any of the edits that have been suggested to you by your peer reviewer, consider having them make a final read-through of your paper to determine if any of the concerns were addressed. If so, move on to the last step. If not, head back and start the review stage again.

Proofreading and Time to Spare

When your paper is complete, it is time for one final proofread. You can read through it yourself, or you can make use of a grammar scanner app on your computer if you want something else to give it one last check. If you do make use of a grammar scanner, keep in mind that they are not infallible—they often do make mistakes, and if you blindly follow the advice of them, you will likely have errors.

Make sure that you do all of this with plenty of time to spare. You should not feel rushed at all when you go through this process, and you should feel like you can, in fact, properly get through everything without a problem. Aim to be done with this stage 2-3 days prior to the due date. This gives you plenty of leeway in case something goes wrong.

5

Considerations

Now, before you get started, there are a few important considerations to keep in mind that will be relevant to making sure that you have everything lined up. In particular, you will want to pay attention to both intellectual property rights and ethical guidelines. You want to comply with these in any good, formal research paper, or you could run into other problems. If you are trying to publish your paper, you will likely run into roadblocks if you do not.

Intellectual Property Rights

Ultimately, intellectual property is anything that is patented, copyrighted, or trademarked. It also encompasses any trade secrets that may be unique to that individual, corporation, or brand, such as the secret to KFC chicken's seasoning. These have legal protections that help to protect those intangible aspects of something; they protect the ideas and inventions, as well as concepts. It can be difficult to keep track of this all, especially because ultimately, there are a lot of different ideas out there—how do you determine which are infringing upon someone else's intellectual property rights?

First, let's take a look at what the three protected concepts that you are likely to encounter during this process are:

- **Patents:**
  This protects inventions, processes, or substances. You may need to keep this in mind if you are trying to, for example, create a version of insulin. The process and the substance would fall under protection by a patent.

- **Trademarks:**
  These protect a product to indicate a producer—they are logos and the like. When you see the trademark, you know what that product is because you recognize it as indicating the brand. Think about the easily recognizable McDonald's arches—they are trademarked.

- **Copyright:**
  This is most likely going to be what you run into—copyright protects the concept of the idea that is expressed, but they do not protect the subject. You may see that, for example, Harry Potter is copyrighted, but the idea of a magical school is not. The idea itself is free from being stolen, but nothing J.K. Rowling could ever do, would create a blanket ban on stories containing those elements. You could go out and make a book called Perry Hotter and have the main character go to Pigboil's School of Wizardry and Witchcraft, and there is not a thing that could be done. You did not take the copyright—you took the same concept, which is not protected. Of course, if you are writing a research paper, you will have to be mindful not to take someone else's thesis and present it as your own—but you can spin-off of it, much in the same way.

Ethical Guidelines for Intellectual Property Rights

When it comes down to it, you must make considerations on how you are going to present and interact with your material. You must be mindful of not infringing upon any of those rights. This means that, in good conscience, you should never make it a point to present something as yours when you have no good claim. Will there be times that what you are putting out there also looks like what someone else is claiming? Probably—there are only so many ideas, after all. However, if you think that you may be infringing upon some of these rights, stop, consider closely, and figure out if you are. Be sensitive—pay attention to what you are doing and your citations.

6

Next Steps: Getting Published

If what you are truly interested in is getting published, you will need to know what that process is like. If you are truly interested in being able to publish research, you have a lot of different options out there for you—you just have to pick out which one works best. Let's take a brief look at what you need to know to get started with your own research publication.

Journal Types

First, you must have an idea of the kind of journal that you want to get published to. There are all sorts of different options out there for you, depending upon the field that you are in. The most common types of articles that you will see include:

- **Letters:**
  These are usually quite short, and they have research findings within them. They are usually designed to push out information immediately without having to publish the entire process. It is essentially the newspaper of academic journals.

- **Research notes:**
  These are quite similar to letters, but are typically less urgent. They will show findings on a subject without going into the full-blown paper writing process.

- **Articles:**
  These usually come between 5 and 20 pages in length and would be the full research paper process that you have seen throughout this whole book.

- **Supplemental articles:**
  These are lists of data that will also summarize up research that is current.

- **Review articles:**
  These are articles that will come up with patterns of review, built up from several articles to create a wider-reaching picture of the subject matter at hand. They are like summaries or wiki websites, where you can find a range of information all provided for you with sources.

The Review Process

When you submit your article to a journal for review, you will usually go through the same process. This can take a range of time based on a particular journal. During this process, an editor screens it and then is peer-reviewed to determine whether it will be published. You can expect to go through the following stages of the review process:

1. You submit your paper to your journal of choice.

2. A journal editor goes through the information to determine what to do with it. Either you will have your manuscript rejected, or you will go on to the next stage.

3. Your paper will then be peer-reviewed, typically without people knowing who wrote it. During this time, between 2 and 6 people will review your journal, and they are typically experts. Your paper will either be accepted by peer reviewers, or they will reject it.

4. The journal editor then determines whether to publish the article or not and when it will be published.

5. You will then be notified of the results.

This process is really quite hands off—you just have to submit your best work to the journal that you are interested in and see what happens.

Conclusion

Congratulations! You have made it to the end of this guide, and hopefully, you are ready to get out there and write your first structured, formal research paper. If all works out well, you should find that your research paper will meet and even potentially exceed expectations. All you have to do is follow the insight that has been carefully crafted for you within this book.

Writing does not have to be a scary process. Your writing process can actually come to be something that you really enjoy doing if you can get the structure down. When you can remove the difficulty from figuring out how to get started and actually begin to write, you may find that the writing process is actually quite enjoyable entirely—you may feel like ultimately, you are thrilled to continue through the process and you are thrilled to be able to spend time researching and writing. Who knows—you may even decide that you will pursue academic research and publication further than just an education, and if you have the skills to back you up, you can be incredibly effective if you do so.

Now, don't be afraid. Arm yourself with the knowledge and the checklist to make sure that your paper is perfectly polished and ready to submit. Get out there and get those fingers pumping out your first paper. It may seem like a lot at first but take your first step. Make a commitment. Write those first 500 words and see where it takes you. You may be pleasantly surprised at the end result.

Do not hesitate—you can start writing proper, formal research papers today if you scroll up right now and start working through this guide with ease.

The hardest part is getting started—and you can start now!

Your Checklist for a Complete Paper

Introduction

- Is the introduction catching and compelling?
- Is the thesis informative and focused?
- Does the thesis make a claim?

Writing format

- Is there direction and continuity?
- Did I include clear transitions?
- Is everything relevant to the thesis?
- Are examples used? Are they properly cited?
- Does the structure make sense?
- Did I provide evidence?
- Does my summary and conclusion reflect my thesis?
- Is it in the past tense and as objective as possible?
- Do my citations match the subject standard?

Research format

- Do I have figures and tables properly assembled, formatted, and titled?
- Did I double-check all data?
- Does everything here make sense?

- Did I clearly state the findings?
- Did I state whether the findings supported my hypothesis?

Proofreading guide

- Did I have at least one other person read it?
- Can I read my paper out loud without running into mistakes?
- Is everything grammatically correct and clear?
- Are there any words that I can cut out that are not 100% necessary?

Final checklist

- Is everything in the proper format? Double-check font, spacing, sizing, etc.
- Is the word count met?
- Did I stay on topic and use enough sources?
- Are all sources cited and listed on their own page?
- Is my paper saved, backed up, and printed out, ready to go?
- Did I add my name to the document?

Thank you

Thank you again for purchasing this book, I hope you have enjoyed it!

If you did enjoy this book, could you leave me a review on Amazon? Just go to your account on Amazon or click on the link below:

CLICK HERE TO LEAVE FEEDBACK ON AMAZON

Thank you so much, it is very much appreciated!

Made in the USA
Monee, IL
19 December 2021